Light

Yes, we can travel faster than light.

Hiroyuki Aizawa

Aizawa Science Museum
Yutaka-cho 1-10-13
Kasukabe, Saitama 3440066, JAPAN
Tel: +81-48-754-9880
e-mail: aizawa@rr.iij4u.or.jp
http://www.aizawa.com/newpage6.html

Preface

"³ God said, "Let there be light," and there was light. ⁴ God saw the light, and saw that it was good. God divided the light from the darkness. ⁵ God called the light "day", and the darkness he called "night". There was evening and there was morning, the first day."
(World Messianic Bible, Genesis)

Once on my high school days, I took a walk coming back home from school with a sunshine around. It was a warm sunny day. When I saw my hands in the gentle sunlight, incidentally one question came up in my mind. "How can I be here safely under the light?" I just learned that light has nature of a particle, which possesses infinite weight when it travels at the speed of light. If it is true, my hands should be crushed by the gentle sunlight. At that time, I did not realize that this incidental and casual question becomes my main theme, which crossed my mind time and time again all through my life.

After finishing high school, I entered the University of Tokyo and got a job after graduation. As a professional scientist, I studied medical life science, especially cell biology. Even during the research, question on light occasionally came to my mind when I observed light from green fluorescent protein under an optical microscope.

On my age at early 40's, I almost finished to solve one of big questions on neuroscience at Johns Hopkins Medical Institute in 2005. And when I spent a resting time at home in Glen Burnie for a while, I incidentally realized a sunlight spot swaying on a wall, which came in my room straight through the chink in the curtain. It was 25 years after my high school days, and still the question came back in my mind as if it was yesterday to think "How can a particle traveling at a speed of light be so gentle for me?"

I came back to Japan in 2006 and finished my professor position at the University of Tokyo last year. Now, I am writing digital books at home for education of kids and young scientists. Maybe I am so old that I could not feel wonder of light anymore. I enjoy having a calm time under the warm sunshine at home as my second life, thinking of future cheerful kids traveling in space of the cosmos faster than the speed of light.

Here, I would like to talk about my short story on "light" for a little bit in this book.

At home in Kasukabe
Mar 6th, 2015
Hiroyuki Aizawa

Preface for the Second Edition

Time flies like an arrow traveling at speed of light since I published the first edition of "Light". After publication of "Light", I wrote more than ten books in English or in Japanese for education of natural science. Among them, I wrote a book entitled "Molecular Mechanics of the Ideal Gas". In the book, I developed a novel model of gas molecules as rotating spherical shell-like elastic bodies contacting with each other. This new model is expected to explain the mechanics of light.

Here, I added one chapter, chapter Ten, in the second edition, in order to develop novel optics based from molecular mechanics of the ideal gas. The optics shown here is quite simple, and readers may feel at ease to understand fundamental aspects of light. Light travels faster in gases under higher pressure because the spherical shell of gas molecules becomes hard for rotation under high pressure.

I hope that this tiny book may kindle young people to explore into space travelling and also that people will organize the solar system under control for development of human society together with living things on earth in the near future.

At home in Kasukabe

Nov 24th, 2016

Hiroyuki Aizawa

Contents

Preface   2

Preface for the Second Edition   4

Contents   5

Chapter 1   Introduction   6

Chapter 2   Electromagnetic Wave Model of Light   9

Chapter 3   Speed of Electromagnetic Wave   11

Chapter 4   Toward the Optics on Earth
in Solar System   12

Chapter 5   Shimmering and Twinkling   14

Chapter 6   Shape of the Rising Sun   16

Chapter 7   Optics on Earth in Solar System   19

Chapter 8   Travelling Faster on Earth than in Space   21

Chapter 9   Air Density and Speed of Light   23

Chapter 10   Molecular Mechanics of Light   25

Chapter 11   Let's Travel Faster than Light!   29

Reference Book   31

Acknowledgements for the First Edition   32

Chapter One
Introduction

In order to understand a fundamental law of light, we are now launching a new exploration of science. How does the sun shine all the planets in solar system? What is the law of nature that produce light at the sun? How does the sunlight travel in the solar system, warm the solar planets, and travel far away out of the Galaxy? Scientists have been approached to these questions for more than five thousand years, and we have acquired a lot of knowledge on light. Based on the scientific data, we'd better to start to discover a secret of light now.

The sun radiates light as sunshine, and the radiation arrays do not appear parallel to each other. When the light shines on the ground, we see the bright world. At the moment, we also feel warm of the light, suggesting that the light has a nature to produce heat on the shined substances. So, it is likely that the light has a calorie. Nowadays, however, phlogiston theory or caloric theory of light has been denied. In the 19th century at Western Europe, a famous physics scientist, Maxwell, revealed that heat is the movement of molecules but not an element because we could produce infinity of heat from a constant mass of irons by friction when digging out a gun barrel. Consequently, it is nonsense to speculate that the light contains phlogiston particles, and Maxwell concluded that light is a wave, which should be defined by his differential equations, rather than an element.   The warmth of the sunshine is nature of light.

Where the sun shines, shadow appears.   Let us watch closely the border of light and shadow.   The border is not a single clear line but a faint outline.   When you watch the shadow of electric cables stringed high from the ground, it is almost impossible to detect the shadow as a line on the road.   The high-voltage line on the steel tower almost completely loses its shadow on the ground.   It has no shadow at all.   This evidence clearly suggests that the light is not a ray travelling straight.   Light is naturally taking a roundabout path.   Moreover, when we put our right thumb and forefinger ultimately close, we can see interference fringes between the two fingers by an eye.   We also see a rainbow pattern when looking at Compact Disk in a view sideways.   These are famous optical phenomena so-called diffraction, indicating that light is a wave.

The wave spreads as the vibration of media.   The sunshine induces vibration of media in space, and spreads to the whole cosmos.   Since a wave is motion of media, the sunshine increases temperature of enlightened substances by inducing the vibration of molecules.

When the sunshine showers on you, you directly feel vibration of sunlight that travelled in space of the solar system.   The speed of light on the earth was measured by Michelson and Morley, and nowadays the speed of light in space is defined as 299,792,458 m/s.

Scientists suggested that light travels in the vacuum rather than among air molecules, which may disturb travelling of light. Consequently, the speed of light is considered to be max at the vacuum in space. The sunlight could shake all the cosmos at the super high speed. Usually in physics, it is well accepted that wave travels much faster in media with high density. For example, sound travels faster in water than in air, and fastest in iron materials. On the other hand, light travels slower in space that contains higher density of molecules. Based on this nature of light, some scientist speculated that light is a particle rather than wave. We will discuss on this issue in detail later.

Anyway, visible sunlight shows a very small wavelength, that is, around 0.5 micrometers. It means the visible sunlight vibrate 10^{15} times per second. This is exactly the beat rate of the sunshine. Since sunlight is a wave, light radiation spread at all the direction equally. The wave strength is getting weaker and weaker when light travels more and more, but never fade out. If the light is a particle, when it radiates far away in space, finally some directions may have no particles while others still have some. This means that when we see the cosmos, someone could see a star at a long distance but others could not. Fortunately, it is not the case for our cosmos as shown later. From the next chapter, we will start scientific travel for asking the mystery of speed of light in much more detail.

Chapter Two
Electromagnetic Wave Hypothesis of Light

Before starting our exploration into a new world of optical science, it is a good idea to go back to a science class of Junior High Scholl for a while to understand the fundamental nature of light.   Let us learn from the past.

When connected to electric battery, a miniature bulb shines brightly.   If you touch the shining bulb, you will find that the bulb is so hot.   Furthermore, if you put a compass around the lead wire connecting the bulb to battery, you will see the compass pointing around the lead.   All these well-known phenomena suggest the relationship among electricity, magnet, heat, and light.   You can produce light and heat from electricity, and you can make electricity from magnet movement.   So, you can produce light, heat, and electricity from magnet movement almost infinitely as revealed by Faraday.

Electricity appears to produce the magnetic field around the lead wire, and magnet movement appears to produce the electric field around the moving magnet.   Maxwell realized that the electromagnetic wave could spread in the field at high speed.

Accordingly, the speed of the electromagnetic wave should depend on the electromagnetic nature of its medium. Scientists studied on the magnetic permeability and permittivity of materials under various conditions, and concluded that the speed of the electromagnetic wave could be comparable to that of light in vacuum. Electric wave produces magnetic wave and vice-versa. Although scientists still argue on the nature of vacuum in space, it was established by Maxwell that electromagnetic wave spreads from the lead wire as fast as light.

Now, it is no doubt that the electromagnetic wave is an action through medium but not at a distance. An amazingly fast speed as light at around 300,000,000 m/s suggests the super rigidity of the medium. Once upon a time, ether was a candidate for the medium of light, but it was rejected by Michelson-Morey experiments. How can it be so rigid as the medium of light? Now scientists believe that the vacuum itself is the medium that has amazing rigidity for travelling at such a super-high speed.

In the next chapter, let us examine the relationship of air density and speed of the electromagnetic wave in detail using Maxwell's electromagnetic equations.

Chapter Three
Speed of Electromagnetic Wave

According to the Maxwell's equations, speed of electromagnetic wave is speculated to become slow when the magnetic permeability and permittivity of the medium increased by molecules in the air. The magnetic permeability and permittivity of air molecules depend on their chemical and physical properties. If the permeability and permittivity of each molecule is kept to be constant under certain conditions, the total permeability and permittivity of the air depends on the air density or the concentration of the molecule in the air. Consequently, the speed of electromagnetic wave should slow down as it comes close to the earth, because the density of gases is higher on the earth than in space. In other words, the air density and speed of electromagnetic wave correlate negatively.

When a wave changes its speed, it is refracted. The atmosphere of the earth functions as an air lens for the electromagnetic wave. By this air lens effect, the shape and motion of subjects existing in space are perturbed when observed from the earth. Since air surrounds the earth as a spherical ball of gas, the lens effect becomes max at the horizontal direction while minimum at the upward direction. It is a good idea to try to apply this electromagnetic hypothesis of light to the optics on earth in the solar system.

Chapter Four
Toward the Optics on Earth in Solar System

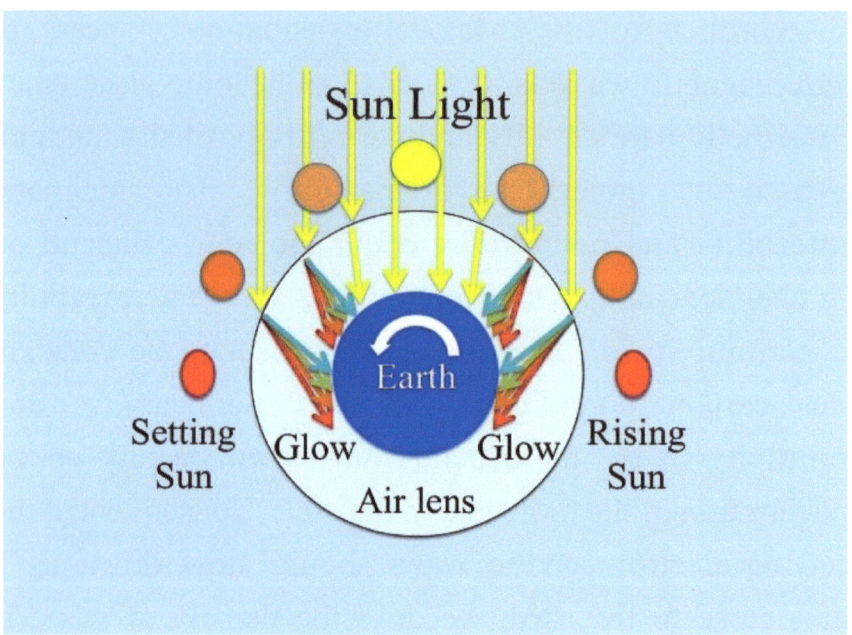

Figure 1.   A working hypothesis of optics on the earth in the solar system by the electromagnetic wave model of light

The light wave travels in space at an extremely high speed at around 300,000,000 m/s in space.   And light is considered to travel slower in the air than in space. Moreover, light travels much slower in a prism than in the air.   Consequently, the sunlight shows refraction when it enters from space into the air.   Sunlight also shows refraction when it enters from the air into a prism.   Since the speed of light in a prism depends on its wavelength, the sunlight is dispersed through a prism. Scientists call this phenomenon as optical dispersion. Light with wavelength from 1 micrometer (infrared) to 0.2 micrometer (ultraviolet) shows dispersion in a prism, causing a beautiful rainbow of sunlight.

In the morning glow, we see the red-colored sun rising from sea. It is so amazing and we feel awe in our mind. Based on the electromagnetic hypothesis of light, the speed of light is considered to decrease when the sunlight enters into the atmosphere on earth. The sunlight is dispersed into rainbow colors by the air lens of the earth as Newton showed using his prism in 17th century at his room. Among the rainbow colors, red light could travel almost in a straight way to present the red sky in the morning glow. This might be the great optics on the earth in the solar system.

When we see the sun from the earth, the morning sun is rising as a red circle from the east, and changing its color from red to a white-yellow gradually. In the afternoon, the sun moves from south to west in the sky. Its color is changing again from white-yellow to red during the setting with glow. This event is a great optical drama of the solar system performed every day, lasting as long as the solar system exists. As discussed in the following chapters, however, this optical model contains a serious problem to be solved.

Chapter Five
Twinkling and Shimmering

When we watch above the sky at night, it is an amazing art of heaven with twinkling stars of Orion, Cassiopeia, and the Big Dipper in the cosmos. All the starlight travels for more than thousands of million years in space.

Twinkling stars do not twinkle by themselves. Stars twinkle because our atmosphere on the earth changes its index of refraction moment by moment. The atmosphere intumesces and shrinks when warmed and cooled, respectively. The speed of light depends on the air density and consequently on the temperature of the atmosphere. Since the atmosphere is a medium of light, swinging of atmosphere density caused flickering of lights. Based on the electromagnetic model of light, we know that the speed of light is faster in low-density or hot air than high-density or cool air.

This mechanism also caused shimmers of hot air or heat hazes. We see a heat haze shimmering above the road shone with strong summer sunlight. We see a heat haze over the hot iron plate of the barbecue. We also see a hot air shimmering on the bonnet of a parking car at seashore. All the phenomena of simmering are caused by the rising hot air draft, which changes refraction index of the atmosphere moment by moment.

All the data obtained above indicate that a spherical

clear ball of high-density gas such as atmosphere on the earth functions as an air lens for optics in solar system.   In the next chapter, we will examine optical nature of the air lens on the earth in detail.   How does the air lens refract the light traveled from the sun to the earth?

Chapter Six
Shape of the Rising Sun

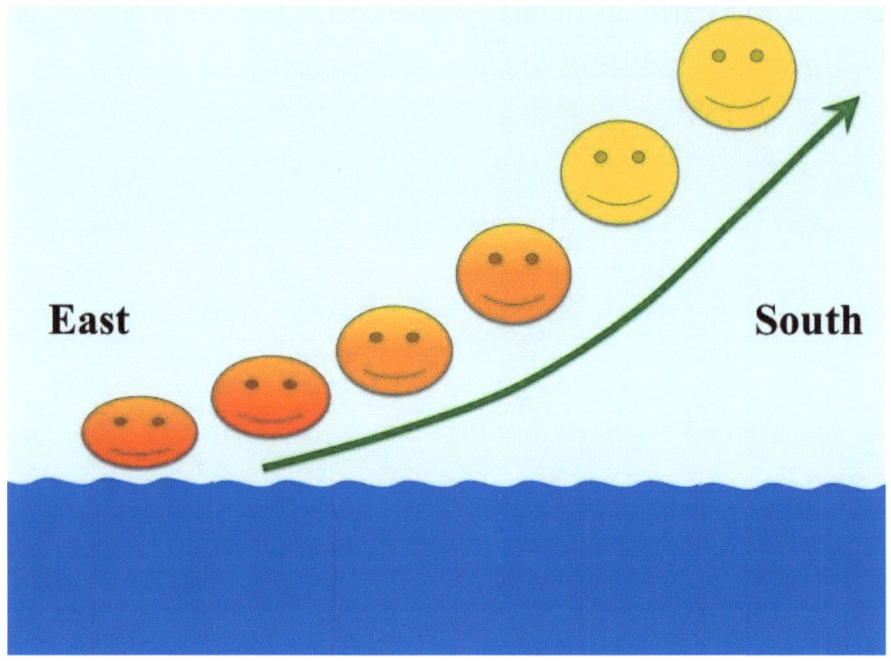

Figure 2.   Changing shape of sun rising from the sea

In the morning glow, we see sun rising in the sky.   In Japan, people climb up Mt. Fuji on Dec 31st to see a rising sun from Pacific Ocean in the morning on the very first day of a Happy New Year.   Since the top of Mt. Fuji is higher than clouds, sometimes we see sun rising from the sea of clouds.   It is a really awesome scene that nature created.

Now, let us see the face and orbit of the rising sun in detail.   The sun will reveal an unexpected and amazing nature of light in the cosmos.

The sun is a perfectly spherical body located at the center of solar system.   How does it look like for us on the earth?   Let us imagine the face of the sun rising from the horizon.   Have you ever seen the rising sun?   If not, you may also imagine the setting sun.   Does it look like a perfect circle?   Or does it look like an egg standing upright on the table?   To our surprise, the rising sun changes its shape moment by moment as illustrated in the Figure 2. Especially just after the dawn in the very early morning, the sun appears red with a vertically short elliptical shape. This optical phenomenon indicates that the speed of light increases when the sunlight get into atmosphere on the earth from outer space.   The precise mechanism of the optical physics of solar system will be discussed in the next chapter.

Let us see the sun rise a little bit more for a while.   The rising sun gradually changes its shape from a vertically short ellipse to a circle moment by moment.   The sun becomes a perfect circle at noon high in the south sky.   The color of the sun also gradually changed from red to yellow according to time or height.   If you take a time-lapse picture of the rising sun, you may also see the track of the sun in the sky. Is the daily track of sun in the sky appears a perfect ring? Surprisingly, the answer is "No".   The track of the sun curves upward as the sun rises from the horizon because of the effect of the air lens around the earth.

We can also realize the air lens effect when we see stars at night.   All stars in the sky rotate around the polar star once a day as the sun does. Probably you have seen an amazing picture of the night sky taken by a long exposure technique, which clearly demonstrate tracks of stars around the polar star. It is exactly the awesome Universe. If you look at the picture in detail, you may realize that any track of stars is not a part of a perfect ring. All the tracks of stars appear as an elliptic shape by the effect of air lens. The Universe tells us a secret of speed of light by the shape of star tracks.

Chapter Seven
Optics on Earth in Solar System

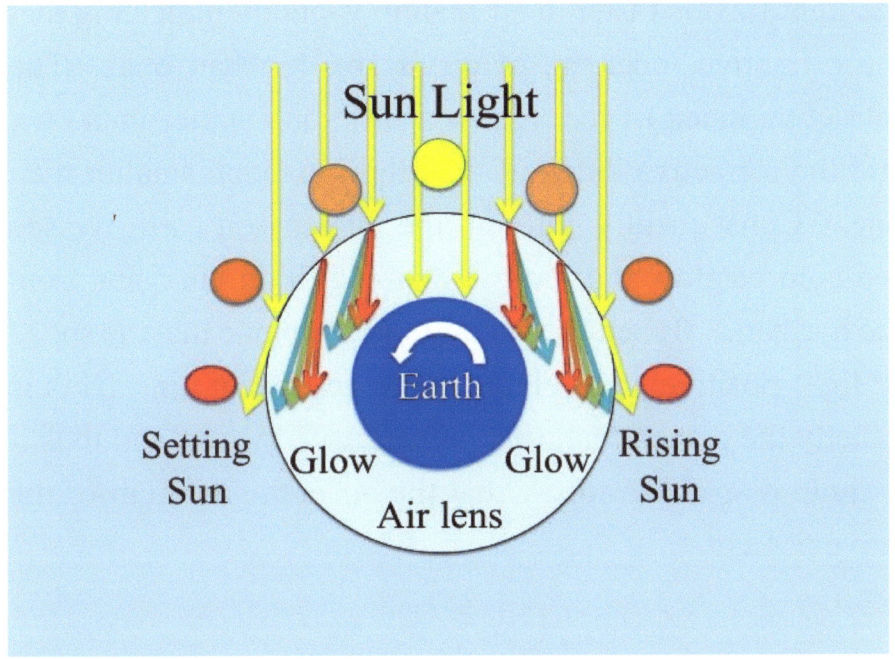

Figure 3.   The optics on the earth in the solar system

After the solar system was created at more than four billion years ago, on every day and night the sun rises from east and sets in west, showing its change of shape along its tracking.

The shape of the spherical sun becomes an ellipse in the morning and evening.   This data indicates that the speed of light increases when the light comes near the earth as shown in Figure 3.   The refractive index of the air is measured by a Michelson's interferometer on the earth.   If light travels slower in air than in vacuum, the refractive index of air is bigger than one. If light travels faster in air than in vacuum, refractive index is smaller than one. Now, we know that the light travels faster in air than in vacuum, indicating that the refractive index, n, of air is smaller than one. The refractive index of red light is a little bit smaller than one, and the refractive index of blue light is much smaller than one. Consequently, through the air lens on earth we see red-colored glow and vertically short elliptic sun rising from the horizon. Taken all together, we conclude that the speed of light positively correlates to the density of air. This is exactly the general nature of a medium of wave, and thus it is quite reasonable to say that the air is the medium of the wave of light.

Chapter Eight
Traveling Faster on Earth than in Space

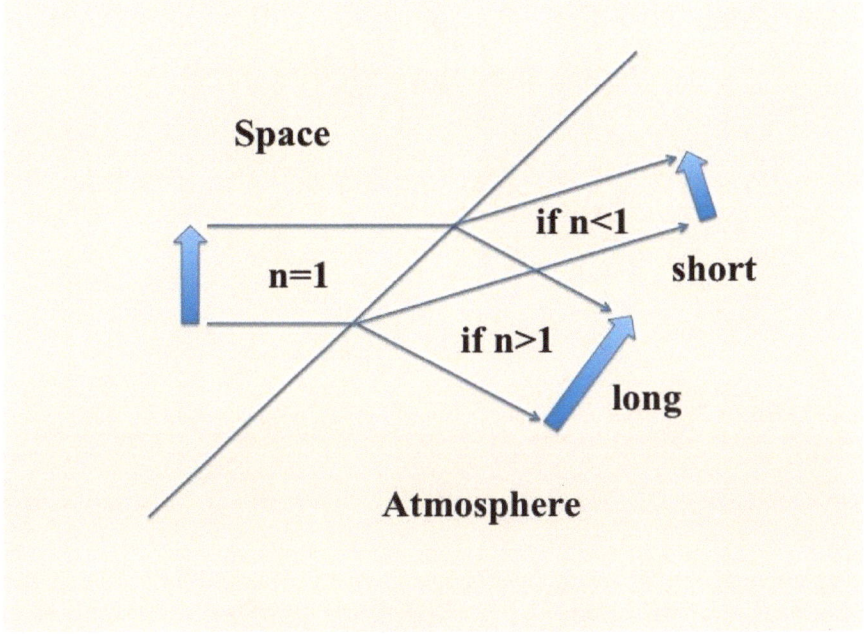

Figure 4. Refraction patterns of light with refractive index of less and more than one

Can we travel faster than the light that travels in space? We already got an answer to this question in the previous chapter. The answer is "Yes". The fact that the refractive index, n, of air is smaller than one indicates that the speed of light in the air is higher than that in space. This is the optical conclusion from the vertically short elliptic shape of the rising sun as shown in the Figure 4. This is a kind of a warp of the reflective index between the space and the atmosphere. Anyway, all of these optical phenomena of light demonstrate typical nature of a wave. Light is a typical wave that travels in a medium or atmosphere faster in higher density medium. There is no wonder anymore on the speed of light. However, we still have a question. What is the nature of a medium of light?

Chapter Nine
Air Density and Speed of Light

We already discussed about the relationship with air density and speed of light in previous chapters. It is a good idea to study on the distribution of the air density along the height on the earth. At first, we should remember the ideal gas law to understand the air density. The law tells us that the air density is proportional to pressure and inversely proportional to temperature. Temperature of atmosphere shifts from 300K on the earth, to 200K at 10 km, 1000K at 300 km, and 3K in space along the vertical height. The sunlight warms the earth, and also the cosmic ray warms some layers of atmosphere depending on the nature of molecules in each layer. The pressure shifts from 1013hPa on earth to 160hPa at 10 km high, and 10^{-5}hPa at 100 km high, and 10^{-13}hPa in space.

These data obviously suggest that the density of the air decreases exponentially and dramatically according to the vertical height. In other word, the density of air is drastically increased when the sunlight enters the stratosphere at around 50 km height from the earth. How thick the atmosphere of earth? While it depends on the definition, the thickness of the atmosphere is around 100 km according to NASA. Since the radius of earth is 6,371 km, the thickness of the atmosphere is less than 2% of the radius of the earth. How thin our air lens is!

As we concluded in the previous chapters, the speed of light increases when the density of its medium or air increases. This is a natural behavior of a wave. Here, we would say that the density of air plays a central role in determination of the speed of light on earth in the solar system. The earth wears a thin air lens whose refractive index is less than one.

Chapter Ten
Molecular Mechanics of Light

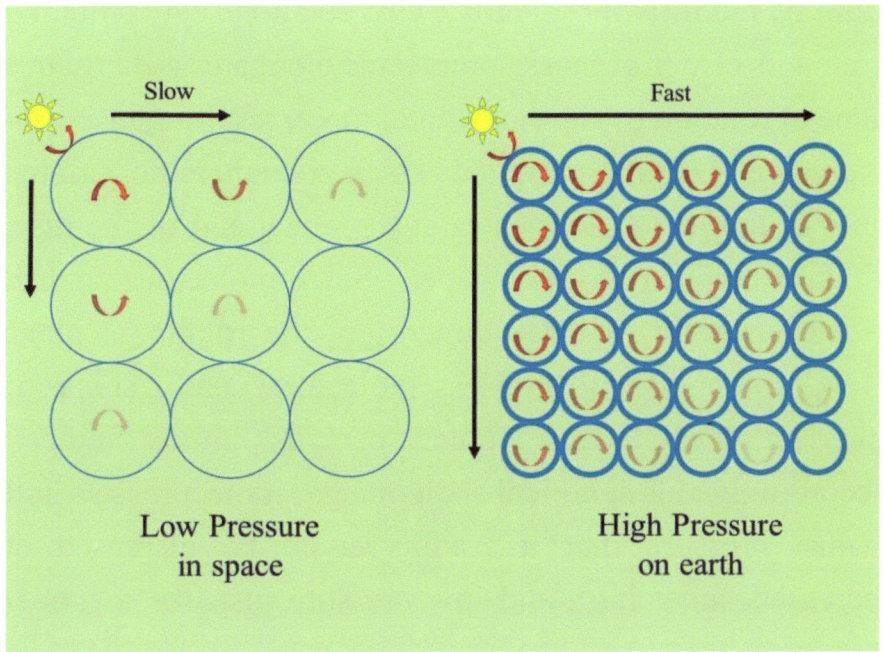

Figure 5. Two-dimensional model of molecular mechanics of light travelling in space (left) or on earth (right). A circle indicates a gas molecule rotating at frequency of light.

If air molecules mediate light, they should contact tightly with each other and must be hard enough for ultra-high speed travelling of light. We know the precise size of molecules such as hydrogen, nitrogen, and oxygen in liquid or solid, which is around 2×10^{-10} m. On the other hand, an averaged diameter of the ideal gas molecule at 273K under 1atm is around 2×10^{-9} m, which is 10-times larger than that of liquid or solid. Thus, it is quite reasonable to speculate that a molecule is an elastic body, whose size could be modified physically by its rotation. As spinning around its center of mass, a molecule elongates radially in a plane of rotation by the centrifugal force, resulting in a disc-like shape. If rotation speed is high enough, almost all the mass of the molecule locates at the distal end of the disc, which becomes ring-like shape. Since molecules contact with each other in three dimension, the rotational plane alters randomly, and thus gas molecules take a shape of spherical shell whose inner part is almost empty. Accordingly, this spherical shell of a gas molecule is so soft against pressure that its radius could be increased or decreased under high and low pressure just like a rubber balloon.

On the other hand, the spherical shell becomes thick and hard for rotation under high pressure. Moreover, the shell could become much harder for rotation till infinity as rotational speed increased under higher pressure. Accordingly, gas molecules could mediate a wave of rotation at so high speed as light under high pressure as shown in Figure 5. Taken all together, it is quite likely that light is a wave of the rotational motion of gas molecules. This is the reason why light travels faster on earth than in space.

As a gas molecule rotates at higher speed under constant

pressure, the molecule becomes larger and consequently its shell becomes thin and soft for rotation. Since temperature is in proportion to square of the rotational speed of a gas molecule, speed of light becomes slower in heat haze than in cool atmosphere under a constant pressure. Accordingly, light is refracted when it passes through heat haze.

Since gas molecules always rotate to create space at around 10^{12} Hz at 300K under 1atm, rotational motion at around or less than 10^{12} Hz could not travel as a wave but diffuses slowly around the molecule as heat. This is quite consistent with the frequency of light observed on the earth from infrared to X-ray (3 x 10^{12} Hz ~ 3 x 10^{18} Hz).

Figure 5 shows two-dimensional model of molecular mechanics of light. The rotation axis of each molecule directs perpendicular to the paper plane, while light travels in the plane, suggesting that the wave is transverse. The rotation vectors of neighboring molecules are antiparallel. In figure 5, the wave of light is shown in two-dimension, and all the molecules rotate in a single plane. This is so-called a "polarized" wave. In three-dimensional space, however, each molecule could have two rotational axes as medium of light.

Taken all together, light is a wave of rotation of gas molecules at frequency higher than that of their stationary rotation. Thus, energy of light corresponds to the rotational energy passing through gas molecules, which is in reverse proportion to square of distance from the origin of light.

Chapter Eleven
Let's Travel Faster than Light!

In the near future, human beings will travel to planets in solar system using a space ship.　We will also travel to stars near the sun such as Alpha Centauri and Sirius.　How fast can we travel in space?　As we discussed above, the speed of light is finite and light travels much slower in space than in the atmosphere on earth.　Light is a wave that travels in the air at limited speed less than 300,000,000 m/s. Consequently, an accelerated body can go faster than light. There is no limit of speed for a physical body to travel in space.　Nowadays, warp drive is just a scientific fiction, but we can definitely make a warp engine in the future. For example, if a wind blows as fast as light somewhere in space, we could not see stars in the wind or behind the wind as they are, because space is warped by the wind optically. Especially, when a space wind blows away from us as fast as light, we could not see anything in the wind and behind the wind just like a black hole.　It should be noted that we do not fly time.　We can go much faster than light in the wind.　We need to prepare for the new period of space ship travelling at the super high speed.

In such a high-speed traveling, we are unable to use the light for telecommunication. During the ultra-light speed driving, we do not see anything behind, and all the stars in front of us shine in blue by Doppler effect.　We need to make a speed meter for the space ship, and also need to make a machine for telecommunication.   What's happen when we go over the speed of light?   Do we see a shock wave of light in space?   Our generation is now standing on the starting line for the warp drive age.   We definitely need to develop an ultra-light speed driving technique to travel outside from solar system for future of human beings.

Who will be a hero in the coming space period?   It's you.   Good luck!

Reference Book

Molecular Mechanics of the Ideal Gas
by Hiroyuki Aizawa
Aizawa Science Museum
2016

Acknowledgements for the First Edition

I finish my tale on speed of light in the last chapter. Yes, we can travel faster than light. I hope that the words will encourage you for the coming cosmic period of human beings. This is a book of light for everybody in the world.

I did not talk on nature of the medium of light so much in this book. Someday somewhere I hope to have an opportunity to tell you about it a little bit more. It is my great pleasure if someone who read this book will start to explore to the new science world and solve the mystery of light. It will be good news for me to hear from you on molecular nature of a medium of light in the near future.

Finally, I would like to express my special thanks to my family, Yoko, Kodai, Arisa, and Hiroto for supporting my life and work to publish this book.

March 11th, 2015
At home in Kasukabe
Hiroyuki Aizawa

Aizawa Science Museum Press
Yutaka-cho, Kasukabe
Saitama, Japan

www.hiroaizawa.com/newpage6.html

www.ingramcontent.com/pod-product-compliance
Lightning Source LLC
Chambersburg PA
CBHW041612180526
45159CB00002BC/819